CHEMISTRY
in the Community

Skill Building Handbook

CHEMCOM

A Project of the American Chemical Society

W. H. FREEMAN AND COMPANY
NEW YORK

Skill Building Handbook

American Chemical Society

Skill Building Handbook Editors: Regis Goode, Patricia Smith

Chief Editor: Henry Heikkinen

Revision Team: Laurie Langdon, Robert Milne, Angela Powers

Revision Assistants: Cassie McClure, Seth Willis

Teacher Edition: Joseph Zisk, Lear Willis

Ancillary Materials: Regis Goode, Mike Clemente

Fourth Edition Editorial Advisory Board: Conrad L. Stanitski (Chair), Boris Berenfeld, Jack Collette, Robert Dayton, Ruth Leonard, Nina I. McClelland, George Miller, Adele Mouakad, Carlo Parravano, Kirk Soulé, Maria Walsh, Sylvia A. Ware (*ex officio*), Henry Heikkinen (*ex officio*)

ACS: Sylvia Ware, Janet Boese, Michael Tinnesand, Guy Belleman, Patti Galvan, Helen Herlocker

W. H. Freeman

Publisher: Michelle Russel Julet

Text Designer: Proof Positive/Farrowlyne Associates, Inc.

Cover Designers: Diana Blume

Illustrations: Proof Positive/Farrowlyne Associates, Inc.

Production Coordinator: Susan Wein

Production Services: Proof Positive/Farrowlyne Associates, Inc.

Supplements and Multimedia Editor: Charlie Van Wagner

Composition: Black Dot Group

Manufacturing: R. R. Donnelley & Sons

Skill building handbook: Chemcom.
    p. cm.
  "A project of the American Chemical Society."
  ISBN 0-7167-3917-8
  1. Environmental chemistry—Study and teaching (Secondary)—Handbooks, manuals, etc. 2. Environmental chemistry—Handbooks, manuals, etc. 3. Environmental chemistry—Problems, exercises, etc. [1. Environmental chemistry—Handbooks, manuals, etc.] I. Title: Chemcom. II. Title: At head of title: Chemistry in the community. III. American Chemical Society.
  TD193 .S58 2000
  540—dc21   00-052070

ISBN 0-7167-3917-8

Library of Congress Cataloging-in-Publication Data

Printed in the United States of America

Second printing, 2002

Contents

Preface

TO THE STUDENT

The material in this *Chemistry in the Community (ChemCom)* **Skill Building Handbook** is designed to help you learn chemistry. It includes information and exercises that support your textbook.

ChemCom is structured around community issues related to chemistry rather than around specific chemical concepts. As a result, chemical concepts are presented on a "need-to-know basis." To fully understand the community issues, you must have a thorough and rich understanding of the chemical concepts involved and know how to apply your learning.

By providing additional explanations, alternative approaches, sample problems, and extra practice, the materials in the **Skill Building Handbook** can help develop and extend your understanding of the chemical concepts. This book will benefit you the most if you exercise your critical thinking skills while you use it. In so doing, you will make the concepts your own.

TO THE TEACHER

The material in this *Skill Building Handbook* falls into two main categories. The first includes material that is not in the *ChemCom* textbook but that students need to successfully learn chemistry. An example of information in this category is the material on density. The National Science Education Standards call for students to understand the characteristic properties of matter, such as density, by grade 8, so we have chosen to leave it out of our student textbook. However, your local science standards may require you to teach density as part of your chemistry course. Thus, we have included information on density in the **Skill Building Handbook** should you need it.

The second category of material includes sections that extend and amplify the concepts presented in the student textbook. To achieve a thorough understanding of science concepts, students often need alternate approaches or **additional** practice. The goal of the **Handbook** is to select topics that perennially give students trouble. In these sections, the teaching approach is different from the one usually found in textbooks. With the variation in student learning styles so commonly found in schools today, a comparable variety in approaches is often the best way to demonstrate that there is more than one approach to solving problems.

Throughout all the *ChemCom* books and ancillaries, the philosophy is to supply the information in the text but let the teacher provide the curriculum appropriate for his or her students. Teaching is an art, involving the coordinated use of appropriate materials and exercises to guide student learning. What is important is that students learn. It is our hope that this **Skill Building Handbook** and the other *Chemistry in the Community* ancillary materials will help make that happen.

Understanding the Uses of Numbers

Chemistry is a quantitative science. Theories are based on and supported by measurements and calculations. Most chemistry experiments involve not only measuring but also a search for the meaning of the measurements. Chemistry students are required to learn how to interpret as well as perform calculations using these measurements.

This part of the skills book provides a brief introduction to understanding the uses of numbers. As you continue your education, you will delve into these topics in greater depth. For now, however, we will review topics you have probably already covered in previous science classes.

MEASUREMENT AND THE METRIC SYSTEM

Metric units were first introduced in France more than 100 years ago. A modernized form of the metric system was internationally adopted in 1960. The system is called "SI," which is an abbreviation of its French name, *Le Système International d'Unités*. SI units are used by scientists in all nations, including the United States. This system has a small number of base units from which all other necessary units are derived.

Quantity	Unit	Abbreviation
Mass	kilogram	kg
Length	meter	m
Time	second	s
Temperature	kelvin	K
Amount of substance	mole	mol
Electric current	ampere	A

Figure 1 *Units of Measurement*

The limited number of units in Figure 1 is not sufficient for every type of measurement a chemist might need to make. For example, the SI unit of length is the meter (symbolized by m). Most doorways are about two meters high. However, many lengths we may wish to measure are either much larger (distance from the earth to the sun) or much smaller (width of a dime) than a meter. To handle such measurements easily, common metric prefixes are used to change the size of the unit. The distance from the earth to the sun can be expressed in kilometers (km), and the width of a dime can be expressed in terms of millimeters (mm). Figure 2 lists the most useful metric prefixes and their meanings.

Prefix	Abbreviation	Meaning	Example
mega-	M	10^6	1 megabyte = 1 000 000 bytes
kilo-	k	10^3	1 kilogram = 1000 grams
deci-	d	10^{-1}	1 deciliter = 0.1 L
centi-	c	10^{-2}	1 centimeter = 0.01 m
milli-	m	10^{-3}	1 milliampere = 0.001 A
micro-	μ	10^{-6}	1 micrometer = 10^{-6} m
nano-	n	10^{-9}	1 nanometer = 10^{-9} m
pico-	p	10^{-12}	1 picometer = 10^{-12} m

Figure 2   *Common Metric Prefixes*

Quantities can be converted from one unit to another through the use of equivalences from Figure 2 and a unit conversion factor.

Example 1: Convert 1456 g to kilograms.

Equivalence: 1 kilogram = 1×10^3 g

From the equivalence we create a **unit conversion factor.** This is a fraction in which the numerator is a quantity that is equal to the quantity of the denominator—except the numerator is expressed in different units. A unit conversion factor is equal to one.

Conversion factor: $\dfrac{1\ kilogram}{1 \times 10^3\ gram}$ or $\dfrac{1 \times 10^3\ gram}{1\ kilogram}$

The conversion factor is chosen to cause the original unit to cancel and the desired unit to remain.

Multiplication: $1456\ g \times \dfrac{1\ kg}{1 \times 10^3\ g} = 1.456\ kg$

More than one conversion factor can be used in a single problem.

Example 2: Convert 325 mg to kilograms.

Equivalences: 1 g = 1×10^3 mg

1 kg = 1×10^3 g

Conversion factors: $\dfrac{1\ g}{1 \times 10^3\ mg}$ *and* $\dfrac{1\ kg}{1 \times 10^3\ g}$

Multiplication: $325 \; mg \times \dfrac{1 \, g}{1 \times 10^3 \; mg} \times \dfrac{1 \, kg}{1 \times 10^3 \, g} \times 3.25 \times 10^{-4} \, kg$

Converting units in this manner is called **dimensional analysis** or **factor labeling.** Dimensional analysis can be used to solve many different types of problems in chemistry. For further instruction on dimensional analysis, see the next section.

Remember that all numbers in chemistry are an outcome of a measurement. As a result numbers should have a measurement unit associated with them. Always include units when you write numbers.

Practice Problems

1. Which metric unit and prefix would be most convenient to measure each of the following?
    a. the diameter of a giant sequoia tree
    b. the diameter of a human hair
    c. time necessary to blink your eye
    d. mass of gasoline in a gallon
    e. mass of a cold virus
    f. amount of aspirin in a tablet
    g. mass of concrete to pave a parking lot

2. What word prefixes are used in the metric system to indicate the following multipliers?
    a. 1×10^3 b. 1×10^{-3} c. 0.01   d. 1×10^{-6}

3. An antacid tablet contains 168 mg of the active ingredient ranitidine hydrochloride. How many grams of the compound are in the tablet?

4. There are 1.609 km in 1 mile. Determine the number of centimeters in a mile.

5. A paper clip is 3.2 cm long. What is its length in millimeters?

6. State at least one advantage of SI units over the customary US units.

Frequently we wish to measure quantities that cannot be expressed using one of the basic SI units. In these situations two or more units are combined to create a new unit. These units are called **derived units.** For example, speed is defined as the ratio of distance to time. To measure speed, two units—distance and time—are combined. Another derived unit frequently used in the laboratory is volume. Let us see how these derived units are related to SI units.

The volume of a cube is determined by multiplying (length $\times$ width $\times$ height). A cube with sides of 10 cm $\times$ 10 cm $\times$ 10 cm has a volume of 1000 cm³, which is defined as a liter.

1000 cm³ = 1 liter (1 L)

Density is the ratio of mass to volume $\left(D = \dfrac{M}{V}\right)$, so it is also a derived unit. It is an important property for determining the identity of a sample of matter. Again, two units are combined—a mass unit and a volume unit. The density of solids and liquids is usually expressed as g/cm³ and the density of gases as g/L.

Practice Problems

1. The average person in the United States uses 340 L of water daily. Convert this to milliliters.

2. A quart is approximately equal to 946 mL. How many liters are in 1 quart?

3. One hundred fifty milliliters of rubbing alcohol has a mass of 120 g. What is the density of rubbing alcohol?

4. A ruby has a mass of 7.5 g and a volume of 1.9 cm^3. What is the density of this ruby?

5. What is the density of isopropyl alcohol if 5.00 mL weigh 3.93 g?

DIMENSIONAL ANALYSIS

Dimensional analysis, also called the factor-label method, is widely used by scientists to solve a wide variety of problems. You have already used this method to convert one type of metric unit to another. The method is helpful in setting up problems and also in checking work because if the unit label is incorrect, the numbers in the answer to the problem are also incorrect. The use of dimensional analysis consists of three basic steps:

1. Identify equivalence relationships in order to create unit conversion factors.

2. Identify the given unit and the new unit desired.

3. Arrange the conversion factor so given units cancel, leaving the new desired unit. Perform the calculation.

The following example illustrates the use of dimensional analysis.

Example 1: In an exercise your laboratory partner measured the length of an object to be 12.2 inches. All other measurements were in centimeters and the answer was to be reported in cm^3. Another member of the group could measure the object in centimeters, or 12.2 inches could be converted to centimeters.

Step 1: Find the equivalence relating centimeters and inches.
$$2.54 \text{ cm} = 1 \text{ in}$$

Step 2: Identify the given unit and the "to find" unit
$$\text{Given unit} = \text{inches}$$
$$\text{"To find" unit} = \text{cm}$$

Step 3: Choose the fraction with the "given" quantity in the denominator and the "to find" quantity in the numerator.
$$12.2 \; \cancel{in} \times \frac{2.54 \; cm}{1 \; \cancel{in}} = 30.0 \; cm$$
Cancel labels as shown above.

Frequently, more than one conversion factor is necessary to solve a particular problem.

Example 2: How many seconds are in 24 hours?

Step 1:   Identify an equivalence:
$$1 \text{ hr} = 60 \text{ min} \qquad 1 \text{ min} = 60 \text{ sec}$$

Step 2:   Given unit: hours; "to find" unit: sec

Step 3:   Arrange for given unit to cancel and progress to the desired unit:
$$24 \text{ hr} \left(\frac{60 \text{ min}}{1 \text{ hr}}\right)\left(\frac{60 \text{ sec}}{1 \text{ min}}\right) = 86\,400 \text{ sec}$$

Practice Problems

1. The distance between New York and San Francisco is 4 741 000 m. Now, that may sound impressive, but to put all those digits on a car odometer is slightly inconvenient. (Of course, in the United States the odometer measures miles, but that is another story.) In this case, kilometers are a better choice for measuring distance. Change the distance to kilometers.

2. Convert 7265 mL to L.

3. The 1500 meter race is sometimes called the "metric mile." Convert 1500 m to miles. (1 m = 39.37 in).

4. The density of aluminum is 2.70 g/cm^3. What is the mass of 235 cm^3 of aluminum?

5. How many 250 mL servings can be poured from a 2.0 L bottle of soft drink?

6. The speed limit in Canada is 100 km/hr. Convert this to meters/second.

7. The density of helium is 0.17 g/L at room temperature. What is the mass of helium in a 5.4 L helium balloon?

8. Liquid bromine has a density of 3.12 g/mL. What volume would 7.5 g of bromine occupy?

9. An irregularly shaped piece of metal has a mass of 147.8 g. It is placed in a graduated cylinder containing 30.0 mL of water. The water level rises to 48.5 mL. What is the density of the metal?

PRECISION AND ACCURACY IN MEASUREMENT

Chemistry experiments often require a number of different measurements, and there is always some error in measurement. How much error depends on several factors, such as the skill of the experimenter, the quality of the instrument, and the design of the experiment. The reliability of the measurement has two components: precision and accuracy. **Precision** refers to how closely measurements of the same quantity agree. A high-precision measurement is one that produces very nearly the same result each time it is measured. **Accuracy** is how well measurements agree with the accepted or true value.

It is possible for a set of measurements to be precise without being accurate. Figure 3 demonstrates different possible combinations of precision and accuracy in an experiment designed to hit the center of the target.

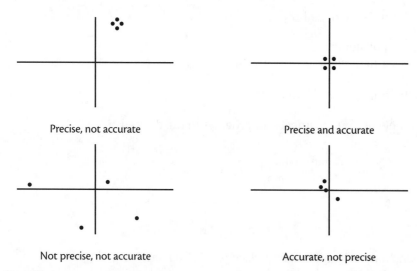

Figure 3 *Precision and Accuracy*

A second example of accuracy and precision is given in Figure 4. The table lists the results of temperature measurements of a beaker of boiling water. The standard temperature of boiling water is 100 °C. The data in the table illustrates the different possible combinations of precision and accuracy in an experiment.

Reading	Thermometer 1	Thermometer 2	Thermometer 3	Thermometer 4
1	99.9 °C	97.5 °C	98.3 °C	97.5 °C
2	100.1 °C	102.3 °C	98.5 °C	99.7 °C
3	100.0 °C	99.7 °C	98.4 °C	96.2 °C
4	99.9 °C	100.9 °C	98.7 °C	94.4 °C
Average	100.0 °C	100.1 °C	98.5 °C	97.0 °C
Range	0.2 °C	4.8 °C	0.4 °C	5.3 °C

Figure 4 *Measured Temperature of 100 mL of Boiling Water*

The average value for each set is taken as the best value. The range—the difference between the largest and smallest values—is the measure of the agreement among the individual measurements. The data taken with Thermometer 1 is accurate and precise, since the average agrees with the accepted value and the range is small. Thermometer 2 provided data that is accurate but not precise since the range is relatively large. The data from Thermometer 3 is precise but not accurate. The range is small enough that it is possible that Thermometer 3 may not have been calibrated properly. Thermometer 4 provides data that is neither precise nor accurate.

Activity

1. Your teacher will provide you with several balances and an object to weigh.

2. Carefully measure the mass of your object 5 times on each balance provided.

3. Calculate the average and the range for each balance. Your teacher will provide you with the "actual" mass of your object.

4. Discuss the accuracy and the precision of each balance you used.

5. Suggest possible errors that might have occurred with the use of each balance.

Practice Problems

1. Determine the precision and accuracy of the following sets of measurements.

   a. A group of students was determining the density of an unknown liquid. They obtained the following values:

   $1.34 \frac{g}{cm^3}$, $1.32 \frac{g}{cm^3}$, $1.36 \frac{g}{cm^3}$. The actual value is $1.34 \frac{g}{cm^3}$.

   b. Another group obtained the same results, but the actual value is $1.40 \frac{g}{cm^3}$.

   c. A third group obtained the following values:

   $1.66 \frac{g}{cm^3}$, $1.28 \frac{g}{cm^3}$, $1.18 \frac{g}{cm^3}$. The actual value is $1.34 \frac{g}{cm^3}$.

   d. A fourth group obtained the following values: $1.60 \frac{g}{cm^3}$, $1.70 \frac{g}{cm^3}$, $1.40 \frac{g}{cm^3}$. The actual value is $1.40 \frac{g}{cm^3}$.

Percent error is a measurement of the accuracy of the measurement. It is calculated using the following formula:

$$Percent\ Error = \frac{Experimental\ value \times Accepted\ value}{Accepted\ value} \times 100\%$$

NOTE: Percent error is a positive number when the experimental value is too high and is a negative number when the experimental value is too low.

SIGNIFICANT FIGURES

As discussed earlier, measurements are an integral part of most chemical experimentation. However, the numerical measurements that result have some inherent uncertainty. This uncertainty is a result of the measurement device as well as the fact that a human being makes the measurement. No measurement is absolutely exact. When you use a piece of laboratory equipment, read and record the measurement to one decimal place beyond the smallest marking on the piece of equipment. The length of the arrow placed along the centimeter stick is 4.75 cm long. There are no graduation markings to help you read the last measurement as 5. This is an estimate. As a result this digit is uncertain. Another person may read this as 4.76 cm. This is acceptable since it is an estimation. There is error (uncertainty) built into each measurement and cannot be avoided.

If the measurement is reported as 4.75 cm, scientists accept the principle that the last digit has an uncertainty of ±0.01 cm. In other words the length might be as small as 4.74 cm or as large as 4.76 cm. It is understood by scientists that the last digit recorded is an estimation and is uncertain. It is important to follow this convention.

Guidelines for Determining Significant Digits

1. **All digits recorded from a laboratory measurement are called significant figures (or digits).**
   The measurement of 4.75 cm has three (3) significant figures.
   NOTE: If you use an electronic piece of equipment, such as a balance, you should record the measurement exactly as it appears on the display.

Measurement	Number of Significant Figures
123 g	3
46.54 mL	4
0.33 cm	2
3 300 000 nm	2
0.033 g	2

Figure 5   *Significant Figures*

2. **All non-zero digits are considered significant.**
   There are special rules for zeros. Zeros in a measurement fall into three types: leading zeros, trailing zeros, and middle zeros.

3. **A middle zero is *always* significant.**

   303 mm: a middle zero—always significant

4. **A leading zero is never significant.** It is only a placeholder; not a part of the actual measurement.

   0.0123 kg: a leading zero—never significant

5. **A trailing zero is significant when it is to the right of a decimal point.**
   This is not a placeholder. It is a part of the actual measurement.

   23.20 mL: a trailing zero—significant to the right of a decimal point

6. **All significant figures include units since they are a result of a measurement.** A number without units has little significance.

> The most common errors concerning significant figures are (1) recording all digits on the calculator readout, (2) failing to include significant trailing zeros (14.150 g), and (3) considering leading zeros to be significant—0.002 g has only one significant figure, not three.

Activity

1. Your teacher will display several different measuring devices. Examine each and determine what digit the last recorded number will occupy.

2. Make the following measurements. Record them using the correct number of significant digits and the measuring device mentioned.

   a. The length of the *ChemCom* book using a meterstick.
   b. The length of the *ChemCom* book using a metric ruler.
   c. The volume of water in a 100-mL graduated cylinder.
   d. The volume of water in a 150-mL beaker.
   e. The volume of water in a 10-mL graduated cylinder.
   f. Measure the mass of a 150-mL beaker.

Practice Problems

1. How many significant figures are in each of the following?

   a. 451 000 m
   b. 6.626×10^{-34} J • s
   c. 0.0065 g
   d. 4056 V
   e. 0.0540 mL

2. For the centimeter rulers below record the length of the arrow shown.

   a.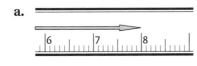

   b.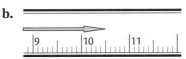

3. The drawings below represent graduated cylinders. Record the volume represented in each picture.

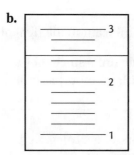

a.

b.

USING SIGNIFICANT FIGURES IN CALCULATIONS

Addition and Subtraction

The number of *decimal places* in the answer should be the same as in the measured quantity with the smallest number of *decimal places*.

$$
\begin{array}{r}
1259.1 \\
2.365 \\
\underline{15.34} \\
1277.075 = \mathbf{1277.1}
\end{array}
$$

Multiplication and Division

The number of *significant figures* in the answer should be the same as in the measured quantity with the smallest number of *significant figures*.

$$
\frac{13.356\,g}{10.42\,mL} = 1.2817658\,g/mL = 1.282\,g/mL
$$

Activity

Find the volume of the block provided by measuring the length, width, and height. Calculate the volume, showing your work, significant digits, and units.

Practice Problems

1. Answer the following problems using the correct number of significant figures.

    a. $16.27 + 0.463 + 32.1$
    b. $42.05 - 3.6$
    c. 15.1×0.032
    d. $13.36/0.0468$
    e. $\dfrac{(13.36 + 0.046) \times 12.6}{1.424}$

2. In the laboratory a group of students was assigned to determine the density of an unknown liquid. They used a buret to measure the liquid

and found a volume of 2.04 mL. The mass was determined on an analytical balance to be 2.260 g. How should they report the density of the liquid?

3. In the first laboratory activity of the year, students were assigned to find the total area of three tabletops in the room. To save time, each of the three students grabbed a ruler and measured the dimensions. They then calculated the area for each tabletop and added them together. Figure 6 presents the students' measurements. What is the total area of the three tabletops?

Student	Length	Width
A	127 cm	74 cm
B	1.3 m	0.8 m
C	50. in	29.5 in

Figure 6 *Tabletop Dimensions*

NOTE: Only numbers resulting from measurements made using instruments have significant figures and have an infinite number of significant figures. Exact numbers include numbers derived from counting and definition. *Examples:* 25 desks in a room or 100 cm = 1 meter

SCIENTIFIC NOTATION

In chemistry we deal with very small and very large numbers. It is awkward to use many zeros to express very large or very small numbers, so scientific notation is used. The number is rewritten as the product of a number between 1 and 10 and an exponential term—10^n, where n is a whole number.

Examples

1. The distance between New York City and San Francisco = 4 741 000 meters:

   4 741 000 m = (4.741 × 1 000 000) m, or **4.741 × 10^6 m**

2. The mass of ranitidine hydrochloride in an antacid tablet = 0.000479 moles:

   0.000479 mol = 4.79 × 0.0001 mol, or **4.79 × 10^{-4} mol**

It is easier to assess the magnitude and to perform operations with numbers written in scientific notation. It is also easier to identify the proper number of significant figures.

Addition/Subtraction Using Scientific Notation

1. Convert the numbers to the same power of ten.

2. Add (subtract) the nonexponential portion of the numbers.

3. The power of ten remains the same.

Example: $(1.00 \times 10^4) + (2.30 \times 10^5)$

A good rule to follow is to express all numbers in the problem to the highest power of ten.

Convert (1.00×10^4) to (0.100×10^5).

$(0.100 \times 10^5) + (2.30 \times 10^5) = 2.40 \times 10^5$

Multiplication Using Scientific Notation

1. The numbers (including decimals) are multiplied.

2. The exponents are added.

3. The answer is converted to scientific notation—the product of a number between 1 and 10 and an exponential term.

Example: $(4.24 \times 10^2) \times (5.78 \times 10^4)$
             $(4.24 \times 5.78) \times (10^{2+4}) = 24.5 \times 10^6$
Convert to scientific notation $= 2.45 \times 10^7$

Division Using Scientific Notation

1. Divide the decimal parts of the number.

2. Subtract the exponents.

3. Express the answer in scientific notation.

Example: $(3.78 \times 10^5) \div (6.2 \times 10^8)$
             $(3.78 \div 6.2) \times (10^{5-8}) = 0.61 \times 10^{-3}$
Convert to scientific notation $= 6.1 \times 10^{-4}$

Calculator Use for Scientific Notation Calculations

Most students use calculators to perform operations with exponential numbers. Scientific calculators have a button labeled ⃞EXP or ⃞EE which enters the "10" portion of the number. Apply the following keystrokes to enter the number 4.741×10^6:

⃞4 ⃞• ⃞7 ⃞4 ⃞1 ⃞EXP ⃞6

If your answers are consistently incorrect by a power of ten, you are probably entering an extra "10" following the ⃞EXP key. When using a calculator to add or subtract exponential numbers, it is not necessary to first convert the numbers to the same power of ten.

Practice Problems

1. Convert the following numbers to exponential notation.

    a. 0.0000369

    b. 0.0452

    c. 4 520 000

    d. $\dfrac{36}{1000}$

    e. 365 000

2. Carry out the following operations:

    a. $(1.62 \times 10^3) + (3.4 \times 10^2)$

    b. $(1.75 \times 10^{-1}) - (4.6 \times 10^{-2})$

    c. $(15.1 \times 10^2) \times (3.2 \times 10^{-2})$

    d. $(6.02 \times 10^{23}) \times (2.0 \times 10^2)$

    e. $(6.02 \times 10^{23}) \div (12.0)$

    f. $\left(\dfrac{6.63 \times 10^{-34} J \cdot s \left(3.00 \dfrac{m}{s} \right)}{4.6 \times 10^{-9}\, m} \right)$

Understanding and Interpreting Graphs and Tables

The ability to interpret graphs and tables is a necessary skill in science but also finds use in everyday life. In articles or textbooks you are likely to find graphs and tables. Understanding the article's message depends heavily on being able to interpret many different types of graphs and tables.

In science tables are used to provide information. Frequently one quantity in a table depends upon or is related to another. Data from tables can be graphed to aid interpretation. Graphs give a visual representation of the data that helps to reveal regularities and patterns.

INTRODUCTORY GRAPHING ACTIVITY

The data table in Figure 7 shows the relationship between the month of the year and the average water temperatures and average dissolved oxygen levels in the Snake River at Riverwood.

Month	Water Temperature (°C)	Dissolved Oxygen (ppm)
January	2	12.7
February	3	12.5
March	7	11.0
April	8	10.6
May	9	10.4
June	11	9.8
July	19	9.2
August	20	9.2
September	19	9.2
October	11	10.6
November	7	11.0
December	7	11.0

Figure 7 *Snake River Data*

Activity

1. Each member of your group will prepare a graph of these data, so you need to make independent decisions about the type of graph you wish to prepare. Be sure to label your axes clearly and to give your graph an informative title. Use only one side of a piece of graph paper.

2. Compare and discuss each group member's graph. List the advantages and disadvantages of the way each graph presents the information provided in the table.

3. Select the graph that the group feels conveys the information in the table most successfully. List the factors that the group used in making this choice.

4. What conclusions about the data can you draw from the graph?

5. As a group, make a list of rules that should be followed in the making of a good graph.

BASIC GRAPHING RULES

1. First decide where the information will be graphed. The horizontal axis (x-axis) is used for the quantity that can be controlled or adjusted. This is called the **independent variable**. The vertical axis (y-axis) is used for the quantity that responds to the changes in the quantity on the x-axis. This is called the **dependent variable**.

2. Choose the scale so the graph becomes large enough to fill most of the available space on the paper.

3. Each regularly spaced division on the graph paper should equal some convenient, constant value. In general, each interval should have a value that can be easily divided visually such as 1, 2, 5, or 10, rather than a value such as 3, 6, 7, or 9.

4. An axis does not need to start at zero, particularly if the plotted values cluster in a narrow range not near zero.

5. Label each axis with the quantity and unit being graphed. For example an axis might be labeled "Temperature, °C."

6. Plot each point. If you plot more than one curve on the same graph, use a different color or geometric shape to distinguish each set of data.

7. For an XY graph, draw a smooth line that lies as close as possible to most of the points. Think of this drawing as a line that is averaging your data. Do not draw a line that connects one point to the next one as in a dot-to-dot drawing. If the curve appears to be straight, draw one continuous line with a ruler.

8. Title your graph with an informative title.

TYPES OF GRAPHS

Graphs are of four basic types: pie charts, bar graphs, line graphs, and XY-plots. The type chosen depends on the characteristics of the data displayed.

Pie Charts

Pie charts show the relationship of parts to a whole. The pie chart in Figure 8 displays the distribution of the world's petroleum reserves. This presentation helps the reader to visualize the magnitude of the differences between various parts of the world. Pie charts are not used as frequently as other types.

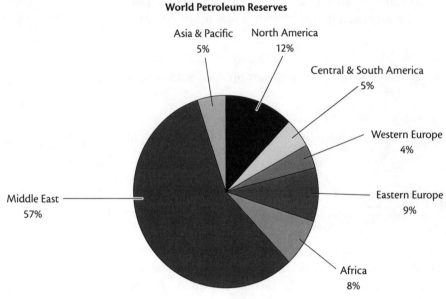

Figure 8   *Sample Pie Chart*

Bar Graphs

Bar graphs and line graphs compare values in a category or between categories. The bar graph in Figure 9 makes a visual comparison of the fat content of types of cheese. This chart might help the viewer choose a cheese snack with a low fat content.

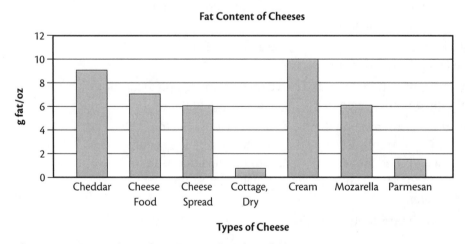

Figure 9   *Comparison of Fat Content in Selected Cheeses*

Bar graphs also can be useful to study trends over time, as in Figure 10. It quickly shows the reader that generally lower temperatures occur in the early morning hours and higher temperatures occur in the late afternoon.

Temperatures also can be predicted for times when a reading was not taken. However, a mathematical relationship between time and temperature is not expected and is not demonstrated. We cannot make any general assumptions for daily temperature graphs that might be constructed at a different location or a different season.

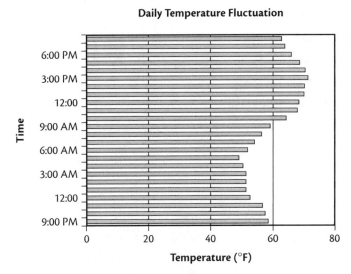

Figure 10 *Temperature Fluctuation Graph*

Multiple Bar Graphs

Multiple bar graphs compare relationships of closely related data sets. Atomic radii plotted against atomic number (Figure 11) show a pattern, but additional interpretation is possible if the elements are divided into periodic table groups as is shown in Figure 12. Some relationships are more easily seen in this format.

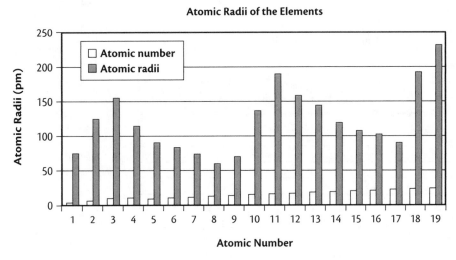

Figure 11 *Atomic Radii*

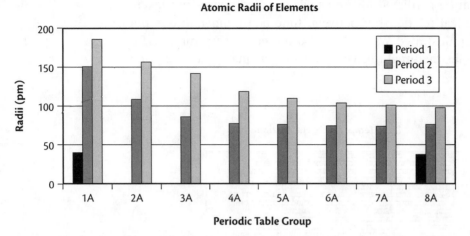

Figure 12   *Periodic Table Group Comparison*

Activity

Answer the following questions based on the graphs in Figures 11 and 12.

1. How does the atomic radius of an element change within a group as the atomic number increases? Is this generalization consistent for all groups illustrated?

2. How does the atomic radius of an element change within a period as atomic number increases? Is this generalization true for all groups illustrated?

Line Graphs

Constructing a **line graph** is another way to show the relationship between two variables. The time and temperature data shown in Figure 10 is probably more easily visualized as a line graph (Figure 13). The same type of information is conveyed.

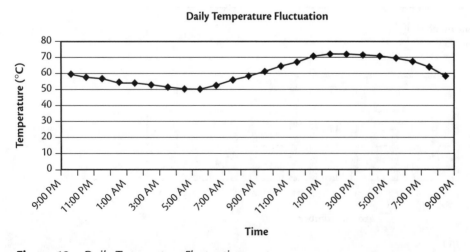

Figure 13   *Daily Temperature Fluctuation*

Activity

Alkanes are compounds of carbon and hydrogen with the general formula, C_nH_{2n+2}. Suppose that you did an experiment to determine the heat of combustion of several alkanes and noticed that the heat of combustion/mole increased as the number of carbons in the alkane increased. The data taken are shown in Figure 14. In your group plot a graph to help interpret the data.

Alkane	Number of carbon atoms	Heat of combustion, kJ/mole
Methane, CH_4	1	891
Ethane, C_2H_6	2	1561
Propane, C_3H_8	3	2219
n-Butane, C_4H_{10}	4	2879
n-Pentane, C_5H_{12}	5	3509

Figure 14 *Increase in Heat of Combustion in Alkanes*

Answer the following questions using your graph.
1. Describe the type of relationship that the graph depicts between carbon atoms and the heat of combustion.
2. Predict the value for the heat of combustion for the alkane with six carbon atoms.
3. Predict the value for the heat of combustion for a substance with no carbon atoms. Why is the value not 0? (*Hint:* Consider what remains in the formula when there are no carbon atoms.)
4. Could you use this same graph to predict the heat of combustion for other kinds of hydrocarbons? Why or why not?

XY-Plots

An **XY-plot** (also called a scatterplot) demonstrates a mathematical relationship between two variables. This type of plot is especially useful in scientific work. Sometimes it is difficult to decide if a graph is a line graph or an XY-plot. One difference is that in an XY-plot it is possible to determine a mathematical relationship between the variables. Sometimes the relationship is the equation for a straight line ($y = mx + b$), but other times it is more complex and requires manipulation of the data. To clarify, we will first look at a straight line, or direct relationship, then proceed to more complex situations.

Example

An entrepreneur was considering investing in a mine that was said to produce gold. Several very small irregular nuggets were given to a chemist for analysis. The chemist, who was instructed to use nondestructive methods, decided to determine the density of the small samples. A micro-buret was used to determine the volume of each nugget, and the mass was determined on an analytical balance. The data collected are shown in Figure 15.

Nugget	Volume (mL)	Mass (g)
1	0.006	0.116
2	0.012	0.251
3	0.015	0.290
4	0.018	0.347
5	0.021	0.386

Figure 15 *Gold Nugget Data*

Because a mathematical relationship is expected between the mass and volume of an element, the chemist constructed an XY-plot.

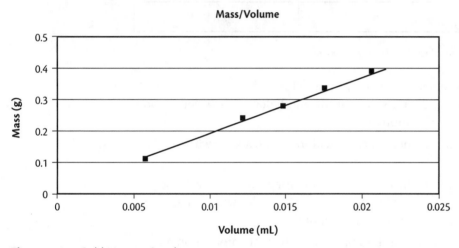

Figure 16 *Gold Nugget Graph*

To connect the plotted points, the best smooth curve, which appears to be a straight line, is drawn.

Activity

1. Find the slope of the line drawn on the Mass/Volume graph. To find the slope, choose two points on the line. These points do not need to be ones you plotted. Determine the x and y coordinates of each point.

    Calculate the slope using the formula $slope = \dfrac{y^2 - y^1}{x^2 - x^1}$

2. What units are appropriate for the slope?

3. The density of a gold sample is 18.88. Is the sample likely to be pure gold? Explain your answer.

Next we will consider a graph where the initial plot is not a straight line. Figure 17 (page 21) provides the data for the graph in Figure 18. It plots the volume of one mole of NH_3 gas at various pressures.

Pressure (atm)	Volume (mL)
0.1000	244.5
0.2000	122.2
0.4000	61.02
0.8000	30.44
2.000	12.17
4.000	5.975
8.000	2.925

Figure 17 *Data for the Effect of Pressure on Volume*

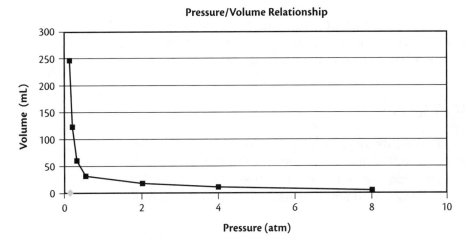

Figure 18 *Graph of Pressure-Volume Relationship*

When the best smooth curve is not a straight line, the data can be manipulated to see if another mathematical relationship is possible. In this case it appears that as the pressure increases the volume decreases. So we can calculate the value of 1/V, add another column to the table (Figure 19), and plot that data (Figure 20).

Pressure (atm)	Volume (mL)	1/Volume, 1/mL
0.1000	244.5	0.00409
0.2000	122.2	0.00816
0.4000	61.02	0.0164
0.8000	30.44	0.0329
2.000	12.17	0.0822
4.000	5.975	0.167
8.000	2.925	0.3419

Figure 19 *Data for Inverse Relationship Between Pressure and Volume*

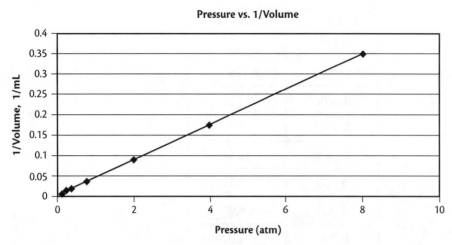

Figure 20 *Graph Showing Inverse Relationship Between Pressure and Volume*

This time the graph exhibits a straight line so we know that pressure and volume are inversely related. If this mathematical manipulation did not result in a straight line, other mathematical changes or analysis might be considered.

INTERPRETING TABLES

Tables can be as simple as listing the value for a single property of a substance or as complex as the one in Figure 21. The unshaded portion lists the melting points for several substances. The shaded portion of the chart suggests some additional information to aid in interpretation. You may be asked to look for relationships in the data.

Substance	Formula	Melting Point (°C)	Molar Mass	Structure	Polarity of Molecule
Water	H_2O	0	18	Molecular	Polar—H bonds
Benzene	C_6H_6	5	78	Molecular	Nonpolar
Naphthalene	$C_{10}H_8$	80	128	Molecular	Nonpolar
Sodium chloride	NaCl	800	58.5	Ionic	Not Applicable
Methane	CH_4	−183	16	Molecular	Nonpolar
Magnesium fluoride	MgF_2	1248	62	Ionic	Not Applicable
Methanol	CH_3OH	−97.8	32	Molecular	Polar—H bonds

Figure 21 *Complex Table Showing Selected Properties of Substances*

Activity

1. Find two compounds in the table with similar molar masses. Compare their melting points. Which of the characteristics listed appears to correlate with the differences in melting point?

2. Compare the molecular compounds with the ionic compounds and make a generalization about structure and melting point.

3. Compare the characteristics of methane, benzene, and naphthalene. What factor seems to be responsible for differences in melting point?

4. The previous three questions use only some of the information available in the table. Write two more questions that might be asked about the table.

5. It is important to use all of the information available in a table. However, you should not make sweeping generalizations that are supported by only a small number of facts. Look at your answer to Question 1 and state what other information you might wish to look up to support your statement.

Additional Problems

1. The graph in Figure 22 shows the approximate level of CO_2 in the atmosphere from 1900 to 1990 for available decades. Study the graph and answer these questions:

   a. Predict the CO_2 levels in 1910, 1940, and 2000.
   b. What other type(s) of graph might also be useful to study this data?

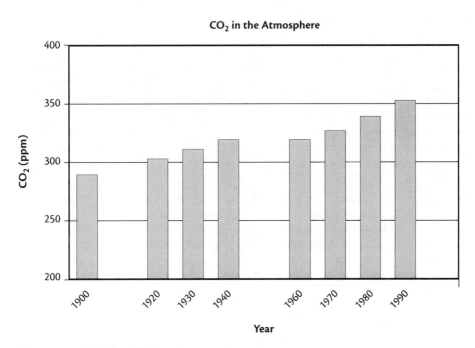

Figure 22   *Relative CO_2 Levels*

2. Graphically determine the density of ethylene glycol for the following data collected in the laboratory.

Volume (mL)	Mass (g)
10.0	11.20
15.0	16.72
20.0	22.14
25.0	17.78
30.0	33.42

Figure 23   *Ethylene Glycol Density Data*

3. The data below was collected when water was heated to its boiling point. Make a plot of this data. Answer the questions.

Time (sec)	Temperature (°C)
0.0	23.0
0.5	27.0
1.0	34.0
1.5	43.0
2.0	58.0
2.5	69.0
3.0	75.0
3.5	83.0
4.0	90.0
4.5	94.0
5.0	96.0
5.5	97.0
6.0	98.0
6.5	99.0
7.0	100.0
7.5	100.5

Figure 24   *Boiling Water Data*

4. Answer the following questions based on your graph.

   a. What type of graph did you choose to plot? Explain why you chose this type.
   b. Describe the change in temperature with time.
   c. Predict the temperature at 4.3 minutes.
   d. Predict the temperature at 8.5 minutes.
   e. During what time period was there the greatest change in temperature?

Understanding Equations and Stoichiometry

Equations are the language of chemistry, and they are important in helping us understand how atoms and molecules form new substances. It has been said many times that chemistry is a quantitative science. Chemistry is also a very precise science. Atoms combine to form molecules in definite ratios. Substances react in definite ratios to form new materials. Atoms and molecules are very small and difficult to count individually, so the idea of a mole was developed.

UNDERSTANDING CHEMICAL EQUATIONS

One way to describe a chemical reaction can be described by writing an English sentence.

> When ethanol burns in the presence of sufficient oxygen, the products are carbon dioxide and water.

Another way is to write a word equation using the chemical names of the reactants and products. The arrow here is often read "yields," but it can also be thought of as an equal sign.

> Ethanol + oxygen → carbon dioxide + water

While a word equation does provide us with additional information, replacing the chemical names with formulas and indicating the physical states of each substance is even more descriptive and specific.

$$C_2H_5OH(l) + O_2(g) \rightarrow CO_2(g) + H_2O(l)$$

However, a chemical sentence is not an equation until it is balanced. The **law of conservation of matter** states that *in a chemical reaction, matter is neither created nor destroyed.* That means the number of atoms of each type must be the same on both sides of the equation. The following chemical equation shows a balanced chemical equation. Count the number of carbon atoms, hydrogen atoms, and oxygen atoms to be sure.

$$C_2H_5OH(l) + 3\,O_2(g) \rightarrow 2\,CO_2(g) + 3\,H_2O(l)$$

Reactant Side	→ Product Side
2 carbon atoms	2 carbon atoms
6 hydrogen atoms	6 hydrogen atoms
7 oxygen atoms	7 oxygen atoms

BALANCING CHEMICAL EQUATIONS

In Unit 2 of your *ChemCom* text you were given instructions to balance equations. You balanced equations by inspecting the number of each type of atom on each side of the equation. You changed coefficients in order to obtain the same number of atoms of each element on the reactant side as on the product side. For some equations this process may be tedious. Presented below is an alternative method for balancing equations. This method uses the law of conservation of matter and simple algebra.

Let's take another look at an unbalanced chemical sentence:

$$\underline{?}\ C_2H_5OH(l) + \underline{?}\ O_2(g) \rightarrow \underline{?}\ CO_2(g) + \underline{?}\ H_2O(l)$$

We know that when this equation is balanced it will have specific values for the coefficient of each reactant and product. We do not know the value of these. For each unknown we can assign a variable:

$$\underline{a}\ C_2H_5OH(l) + \underline{b}\ O_2(g) \rightarrow \underline{c}\ CO_2(g) + \underline{d}\ H_2O(l)$$

We can now write an algebraic expression for each element in this reaction. When the equation is balanced, the number of carbon atoms on the reactant side must be equal to the number of carbon atoms on the product side (law of conservation of matter).

On the reactant side there are $2a$ carbon atoms (we multiply the coefficient by the subscript). On the product side there are $1c$, or c, carbon atoms, so we can say:

Carbon atoms: $2a = c$

Repeating this process for each element in the reaction we obtain the following:

Hydrogen atoms: $6a = 2d$

Oxygen atoms: $a + 2b = 2c + d$

At this point we assign one of the variables the value of 1 and solve for the remaining variables. Generally it is best to assign the value of 1 to the variable for the most complicated compound. Since a is assigned to C_2H_5OH, choose $a = 1$.

So if $a = 1$, then $2(1) = c$ and $c = 2$.

And $6(1) = 2d$, so $d = 3$

Finally $1 + 2b = 2(2) + 3$,

so when you solve for b, $b = 3$

Substitute the numerical values into the chemical equation and count atoms to ensure that the equation is balanced.

$$\underline{1}\ C_2H_5OH(l) + \underline{3}\ O_2(g) \rightarrow \underline{2}\ CO_2(g) + \underline{3}\ H_2O(l)$$

Example 1

Balance the following equation: $SiO_2(s) + HF(aq) \rightarrow SiF_4(g) + H_2O(l)$

Step 1: Assign variables for coefficients for each reactant and product:

$\underline{a}\ SiO_2(s) + \underline{b}\ HF(aq) \rightarrow \underline{c}\ SiF_4(g) + \underline{d}\ H_2O(l)$

Step 2: Write algebraic equations for the number of each element on each side of the equation:

Silicon atoms: $a = c$

Oxygen atoms: $2a = d$

Hydrogen atoms: $b = 2d$

Fluorine atoms: $b = 4c$

Step 3: Assign one of the variables the value of 1:

$a = 1$

Step 4: Solve for the other variables using the assignment made in Step 3.

If $a = 1$, then $c = 1$,

and $d = 2$,

and $b = 4$.

Step 5: Substitute these values into the equation and check:

$\underline{1}\ SiO_2(s) + \underline{4}\ HF(aq) \rightarrow \underline{1}\ SiF_4(g) + \underline{2}\ H_2O(l)$

(**Note:** The coefficient that has a value of 1 is understood and is generally not written. It was written in the previous examples only for emphasis.)

Example 2

Balance the following equation: $C_2H_6(g) + O_2(g) \rightarrow CO_2(g) + H_2O(g)$

Step 1:

$\underline{a}\ C_2H_6(g) + \underline{b}\ O_2(g) \rightarrow \underline{c}\ CO_2(g) + \underline{d}\ H_2O(g)$

Step 2:

Carbon atoms: $2a = c$

Hydrogen atoms: $6a = 2d$

Oxygen atoms: $2b = 2c + d$

Step 3:

$a = 1$

Step 4:

If $a = 1$, then $c = 2$,

$d = 3$,

$b = 7/2$.

Occasionally we obtain a fraction as a solution. If this occurs we multiply all the variables by a common factor to eliminate fractions.

$a = 1 \times 2 = 2$

$c = 2 \times 2 = 4$

$d = 3 \times 2 = 6$

$b = 7/2 \times 2 = 7$

Step 5:

$$\underline{2}\ C_2H_6(g) + \underline{7}\ O_2(g) \rightarrow \underline{4}\ CO_2(g) + \underline{6}\ H_2O(g)$$

Practice Problems

Balance the following equations.

1. $Zn + HCl \rightarrow ZnCl_2 + H_2$
2. $Al + O_2 \rightarrow Al_2O_3$
3. $C_4H_{10} + O_2 \rightarrow CO_2 + H_2O$
4. $KClO_3 \rightarrow KCl + O_2$
5. $Fe + H_2O \rightarrow Fe_3O_4 + H_2$
6. $CaC_2 + H_2O \rightarrow C_2H_2 + Ca(OH)_2$
7. $MnO_2 + HCl \rightarrow MnCl_2 + H_2O + Cl_2$
8. $Fe_2O_3 + CO \rightarrow Fe + CO_2$
9. $H_2O_2 \rightarrow H_2O + O_2$
10. $C_2H_5OH + O_2 \rightarrow CO_2 + H_2O$

Write balanced chemical equations for the reactions that occur in each of the following situations.

11. When benzene (C_6H_6) combines with the oxygen (O_2), the products are carbon dioxide (CO_2) and water (H_2O).

12. When photosynthesis takes place in a green plant, carbon dioxide and water combine to produce glucose ($C_6H_{12}O_6$) and oxygen.

13. Nitroglycerin ($C_3H_5N_3O_9$), a drug used for heart pain problems, is synthesized from glycerin ($C_3H_8O_3$) and nitric acid (HNO_3) in the presence of a catalyst. Water is also a product of the reaction.

14. Some antacids contain aluminum hydroxide [$Al(OH)_3$], which reacts with the acid (HCl) in the stomach to produce aluminum chloride ($AlCl_3$) and water.

15. An antacid remedy contains sodium bicarbonate ($NaHCO_3$) and citric acid ($H_3C_6H_5O_7$), which react to produce carbon dioxide (the source of the familiar fizz), sodium citrate ($Na_3C_6H_5O_7$), and water.

16. When table sugar ($C_{12}H_{22}O_{11}$) is heated, water vapor and elemental carbon form.

WHAT IS A MOLE?

The term *mole* is classified as a counting number, a number used to specify a certain number of objects. Pair and dozen are other examples of counting numbers. Seldom does anyone purchase computer paper by the sheet. Instead you buy a ream, or package of 500 sheets. At the grocery store you buy eggs by the dozen. Many other objects are identified in packages of this size—

rolls, ears of corn, file folders. When you buy a package of a dozen, you know you will get twelve objects. A mole equals 6.02×10^{23} objects! Most often things that are counted in units of moles are very small—atoms, molecules, or electrons. Because it would take an impossibly long time to count 6.02×10^{23} objects, an indirect method is used. An analogy may be helpful. Sometimes people save money by tossing all coins of a certain value into a large container. When the container is full they take it to the bank. Instead of counting each coin, the bank employee finds the mass of all of the coins and divides the result by the mass of one coin. (How long do you think it would take to count a mole of dimes?)

Activity: Weighing Objects to Determine Their Number

1. Your teacher will give you a container of objects. Weigh one of them.

2. Empty the container of objects into the 250-mL beaker. Determine the mass of the beaker and the objects.

3. Subtract the mass of the beaker from the total mass.

4. Calculate the number of objects in the beaker by dividing the total mass of objects by the average mass of one object.

5. Count the number of objects in the beaker to determine if your calculation is correct. Explain why your calculation was correct or incorrect.

MOLES AND MOLAR MASS

The modern definition of a mole is the number of atoms in exactly 12 grams of the carbon-12 (C-12) isotope. This number itself is named after Amedeo Avogadro, who investigated related concepts but never determined the number. At least four different types of experiments have determined that the number is 6.02×10^{23}. Avogadro's number is known to ten significant figures, but three will be enough for most of your calculations.

$$1 \text{ mole} = 6.022126736 \times 10^{23} \text{ particles}$$

The modern atomic-weight scale is also based on C-12. The relative mass of a hydrogen atom compared to a carbon atom is 1.008. Therefore, one mole of hydrogen atoms has a mass of 1.008 g, and one mole of oxygen atoms is 15.9994 g. The number of grams in one mole, or **molar mass,** of a compound is found by adding the relative atomic masses of the atoms in the formula.

Example: Determine the molar mass of water, H_2O.

$$2 \text{ mol H} \times \frac{1.01 \text{ g H}}{1 \text{ mol H}} = 2.0 \text{ g H}$$

$$1 \text{ mol O} \times \frac{16.00 \text{ g O}}{1 \text{ mol O}} = 16.0 \text{ g O}$$

Molar mass of water   = 18.0 grams

Practice Problems

Find the molar mass (grams in one mole) of each of the following:

1. Acetic acid, CH_3COOH

2. Formaldehyde, $HCHO$

3. 2-Dodeconol, $CH_3(CH_2)_9CH(OH)CH_3$

4. Glucose, $C_6H_{12}O_6$

5. Ethanol, C_2H_5OH

6. Phosphoric acid, H_3PO_4

7. Cobalt(II) chloride hexahydrate, $CoCl_2 \cdot 6\ H_2O$

MASS-MOLE CONVERSIONS

Conversions between masses and moles can best be accomplished by using dimensional analysis.

Example

What mass (in grams) of water contains 0.25 mol H_2O?

The mass of one mole of water (18.0 g) is found as shown above. The two common factors are

$$\frac{1\ mol\ H_2O}{18.0\ g\ H_2O} \quad and \quad \frac{18.0\ g\ H_2O}{1\ mol\ H_2O}$$

The second conversion factor is chosen in order to cancel the labels. Units for the answer are g H_2O, as expected.

$$0.25\ \cancel{mol\ H_2O} \times \frac{18.02\ g\ H_2O}{1\ \cancel{mol\ H_2O}} = 4.5\ g\ H_2O$$

How many moles of water are present in 1.00-kg sample of water?

$$1.00\ kg\ H_2O \times \frac{1000\ g\ H_2O}{1\ kg\ H_2O} \times \frac{1\ mol\ H_2O}{18.0\ g\ H_2O} = 55.5\ mol\ \ H_2O$$

Practice Problems

1. Acetic acid, CH_3COOH, and salicylic acid, $C_7H_6O_3$, combine to form aspirin. If a chemist uses 5.00 g salicylic acid and 10.53 g acetic acid, calculate the number of moles of each compound used.

2. Dodecanol, $CH_3(CH_2)_9CH(OH)CH_3$ is used in synthesis of wetting agents. A manufacturer orders 500.0 kg of the compound. How many moles are ordered?

3. Calcium chloride hexahydrate, $CaCl_2 \cdot 6\ H_2O$, is sprinkled on sidewalks to melt ice and snow. How many moles of the compound are in a 5.0 kg sack of the material?

4. 1.5 mol sodium hydroxide, $NaOH$, are required to prepare a solution. What is the equivalent number of grams?

5. The laboratory technician must prepare a solution that requires 0.123 mol silver nitrate, $AgNO_3$. How many grams are necessary?

A QUANTITATIVE UNDERSTANDING OF FORMULAS

Calculating Percent Composition from a Formula

The percent by mass of each material found in an item is called the element's **percent composition.** To find the percent composition of an element in a compound, first calculate the compound's molar mass. Then divide the total mass of the element in the compound by the molar mass of the compound and multiply the result by 100%.

Example

Calculate the molar mass of sucrose, $C_{12}H_{22}O_{11}$.

$$12 \text{ mol C } (12.01 \text{ g/mol}) = 144.1 \text{ g}$$

$$22 \text{ mol H } (1.01 \text{ g/mol}) = 22.2 \text{ g}$$

$$\underline{11 \text{ mol O } (16.0 \text{ g/mol}) = 176.0 \text{ g}}$$

Total: Molar mass of $C_{12}H_{22}O_{11}$ = 342.3 g

Find the mass % of each element in sucrose:

$$\% \text{ C} = \frac{\text{mass C/mol}}{\text{mass sucrose/mol}} \times 100 = \frac{144.0 \text{ g C/mol}}{342.0 \text{ g sucrose/mol}} \times 100\% = 42.11\% \text{ C}$$

$$\% \text{ H} = \frac{\text{mass H/mol}}{\text{mass sucrose/mol}} \times 100 = \frac{22.2 \text{ g H/mol}}{342.2 \text{ g sucrose/mol}} \times 100\% = 6.48\% \text{ H}$$

$$\% \text{ O} = \frac{\text{mass O/mol}}{\text{mass sucrose/mol}} \times 100 = \frac{176.0 \text{ g O/mol}}{342.2 \text{ g sucrose/mol}} \times 100\% = 51.43\% \text{ O}$$

Practice Problems

1. Find the percent carbon and percent hydrogen in C_2H_4.
2. Find the percent carbon and the percent hydrogen in C_3H_6.
3. What do the answers to problems 1 and 2 have in common? Explain your answer.
4. Find the percent of each element in the following compounds.
   a. $MgSO_4$
   b. $MgSO_4 \cdot 7 H_2O$
   c. $(CH_3)_2SO$
   d. C_5H_5N
   e. $Ca_3(PO_4)_2$

Calculating Percent Composition from Mass

In the laboratory it is possible to determine the percentage composition for many compounds.

Example

In the laboratory 0.400 g magnesium ribbon is burned in the presence of oxygen to form magnesium oxide. The product has a mass of 0.663 g. What is the percentage composition of the compound?

Mass of magnesium: 0.400 g

Mass of magnesium oxide: 0.663 g

$$\% \, Mg = \frac{0.400 \text{ g Mg}}{0.663 \text{ g compound}} \times 100 = 60.3\% \text{ Mg}$$

$$\% \, O = 100\% - 60.3\% = 39.7\% \text{ O}$$

Practice Problems

1. A compound consists of 11.66 g Fe and 5.01 g O. What is the percent of each element in the compound?

2. A sample of adrenaline, an important hormone, is made up of 0.590 g carbon, 0.071 g hydrogen, 0.262 g oxygen, and 0.077 g nitrogen. What is the percent of each element in the compound?

3. Copper (II) sulfate commonly appears as a hydrated salt. In the lab a sample of hydrated salt was heated until all the water was driven off. The original sample had a mass of 24.50 g. The dried sample has a mass of 15.66 g. What is the percent of water in this compound?

Formulas from Percentage Composition: Empirical Formulas

An **empirical formula** gives the relative numbers of different elements in a substance, using the smallest whole numbers for subscripts. The empirical formula of the compound can be calculated from percent composition data or from the number of grams of each element in the sample. A formula compares the relative numbers of moles of each element, so if percents are used, they must be converted to grams, and grams to moles. It is easiest to assume there is 100 g of the compound. Then the percentages are equal to the number of grams of each element.

Example 1

A hydrocarbon (the only elements in the compound are hydrogen and carbon) consists of 85.7% carbon and 14.3% hydrogen. What is its empirical formula?

Step 1:   Assume that you have 100 g of a compound and that 85.7 g of it are carbon and 14.3 g are hydrogen.

Step 2:   Use dimensional analysis to find the number of moles of
each element in the compound.

$$85.7 \text{ g C} \times \frac{1 \text{ mol C}}{12.0 \text{ g C}} = 7.14 \text{ mol C}$$

$$14.3 \text{ g H} \times \frac{1 \text{ mol H}}{1.01 \text{ g H}} = 14.2 \text{ mol H}$$

Step 3:   Determine the smallest whole number ratio of moles of
elements by dividing all mole values by the smallest.

$$\frac{7.14 \text{ mol C}}{7.14} = 1 \text{ mol C}$$

$$\frac{14.2 \text{ mol H}}{7.14} = 1.99 \text{ mol H} \approx 2 \text{ mol H}$$

**The ratio of moles of C to moles of H = 1:2 and the empirical formula
for the compound is CH_2.**

Example 2

The percentage composition of one of the oxides of nitrogen is 74.07%
oxygen and 25.93% nitrogen. What is the empirical formula?

Step 1:   100 g of compound consists of 74.07 g oxygen and 25.93 g
nitrogen.

$$74.07 \text{ g O} \times \frac{1 \text{ mol O}}{16.00 \text{ g O}} = 4.62 \text{ mol O}$$

$$25.93 \text{ g N} \times \frac{1 \text{ mol N}}{14.01 \text{ g N}} = 1.85 \text{ mol N}$$

Step 2:

$$\frac{4.62 \text{ mol O}}{1.85} = 2.5 \text{ mol O}$$

Step 3:

$$\frac{1.85 \text{ mol N}}{1.85} = 1.0 \text{ mol N}$$

Step 4:   This ratio does not consist only of whole numbers, but formulas are not expressed with decimals: $NO_{2.5}$. Multiply all the
numbers in the ratio by a number that converts the
decimal to a whole number. In this case the number is 2, and
the empirical formula becomes N_2O_5.

Practice Problems

1. A compound is 82.3% nitrogen and 17.6% hydrogen. What is the empirical formula?

2. Another compound is 52.2% C, 13.0% H, and 34.8% O. What is the
   empirical formula?

3. Nicotine is 74.03% C, 8.70% H, and 17.27% N. What is the empirical
   formula?

4. A hydrocarbon is found to be 82.66% carbon. What is the empirical formula?

5. A sample of a compound contains 0.252 g titanium and 0.748 g chlorine. Determine the empirical formula of this compound.

UNDERSTANDING MASS RELATIONSHIPS IN CHEMICAL REACTIONS

If the number of atoms is conserved in a chemical reaction, the mass must also be conserved as expected from the Law of Conservation of Mass. In the equation for the formation of water—$2 H_2(g) + O_2(g) \rightarrow 2 H_2O(l)$—2 molecules of hydrogen and 1 molecule of oxygen combine to form 2 molecules of water. We could also say that 2 moles of hydrogen react with 1 mole of oxygen to form 2 moles of water. Using the number of grams in a mole of each substance, the mass relationships in the table can be determined. The ratio of moles of hydrogen to moles of oxygen to form water will be 2:1. If 10 moles of hydrogen are available, 5 moles of oxygen are required.

$2 H_2(g)$	+	$O_2(g)$	$\rightarrow$	$2 H_2O(l)$
2 molecules		1 molecule		2 molecules
2 mol		1 mol		2 mol
2 mol $\times$ (2.02 g/mol)		1 mol $\times$ (32.00 g/mol)		2 mol $\times$ (18.02 g/mol)
4.04 g		32.00 g		36.04 g

Solving problems involving the masses of products and or reactants is conveniently accomplished by dimensional analysis. All numerical problems involving chemical reactions begin with a balanced equation.

Example

Find the mass of water produced when 10.0 grams hydrogen react with plenty of oxygen.

$$2 H_2(g) + O_2(g) \rightarrow 2 H_2O(l)$$

Step 1: Find the moles of hydrogen represented by 10.0 g by using the molar mass of H_2.

$$10.0 \text{ g } H_2 \times \frac{1 \text{ mol } H_2}{2.02 \text{ g } H_2} = 4.95 \text{ mol } H_2$$

Step 2: Find the number of moles of H_2O produced by 4.95 mole H_2. From the balanced equation, we know that for every 2 mol H_2 we produce 2 mol H_2O.

$$4.95 \text{ mol } H_2 \times \frac{2 \text{ mol } H_2O}{2 \text{ mol } H_2} = 4.95 \text{ mol } H_2O$$

Step 3:   Find the mass of H_2O that contains 4.95 mol H_2O by using the molar mass of water.

$$4.95 \text{ mol } H_2O \times \frac{18.0 \text{ g } H_2O}{1 \text{ mol } H_2O} = 89.1 \text{ g } H_2O$$

Most chemistry students find it is more convenient to set up all three steps in one problem. Make sure that all labels cancel except the g H_2O, an appropriate unit to express the mass of H_2O as required in the problem.

$$10.0 \text{ g } H_2 \times \underbrace{\frac{1 \text{ mol } H_2}{2.02 \text{ g } H_2}}_{\substack{\text{Molar mass} \\ \text{of } H_2}} \times \underbrace{\frac{2 \text{ mol } H_2O}{2 \text{ mol } H_2}}_{\substack{\text{Coefficients} \\ \text{in equation}}} \times \underbrace{\frac{18.0 \text{ g } H_2O}{1 \text{ mol } H_2O}}_{\substack{\text{Molar mass} \\ \text{of } H_2O}} = 89.1 \text{ g } H_2O$$

Practice Problems

1. Acetylene burns in air to form carbon dioxide and water:

   $$5 \text{ C}_2H_2(g) + 2 \text{ O}_2(g) \rightarrow 4 \text{ CO}_2(g) + 2 \text{ H}_2O(l)$$

   How many moles of CO_2 are formed from 25.0 moles C_2H_2?

2. If insufficient oxygen is available, carbon monoxide can be a product of the combustion of butane: $9 \text{ C}_4H_{10}(l) + 2 \text{ O}_2(g) \rightarrow 8 \text{ CO}(g) + 10 \text{ H}_2O(l)$. What mass of CO could be produced from 5.0 g butane?

3. 15.0 g $NaNH_2$ is required for an experiment. Using the following reaction, what mass of sodium metal is required to produce the $NaNH_2$?

   $$2 \text{ Na}(s) + 2 \text{ NH}_3(g) \rightarrow 2 \text{ NaNH}_2(s) + \text{H}_2(g)?$$

4. Ethanol and acetic acid react to produce ethyl acetate according to the reaction $C_2H_5OH + CH_3COOH \rightarrow CH_3COOC_2H_5$. If the reaction is only 35% efficient at the conditions used, what mass of CH_3COOH will be necessary to produce 100 g $CH_3COOC_2H_5$? Assume that sufficient ethanol is available.

5. Heating $CaCO_3$ yields CaO and CO_2. Write the balanced equation. Calculate the mass of $CaCO_3$ consumed when 4.65 g of CaO forms.

UNDERSTANDING LIMITING REACTANTS

In Unit 4, Section B of your *ChemCom* text you will find a conceptual atom-counting method for finding the limiting reactant of an equation. Refer to that discussion before you begin this presentation, which presents two additional methods for finding limiting reactants.

If a combustion problem states that excess oxygen is available, we need not concern ourselves with the oxygen-to-water ratio. The mass of water produced is predicted from (and limited by) the mass of hydrogen available. Of course in the laboratory we often deal with specified masses of each reactant, but that requires an enhanced problem-solving method.

Required Ingredients	Available Ingredients	Possible Casseroles
1 27-oz can whole green chiles	2 27-oz cans whole green chiles	2
1 lb Monterrey Jack cheese	3 lb Monterrey Jack cheese	3
1 lb cheddar cheese	3 lb cheddar cheese	3
3 eggs	1 doz eggs	4
3 Tbsp flour	5 lb flour	Many
5 oz canned evaporated milk	4 5-oz cans evaporated milk	4

Figure 26   *Required Ingredients Table*

How many casseroles can be made? Although four casseroles can be made from the available eggs or milk, there are only enough cans of green chiles for two casseroles. In other words, the number of cans of green chiles can be called the limiting factor. After the two casseroles are prepared, cheese, eggs, flour, and milk will remain, but all the green chiles will be used. Therefore, no more casseroles can be made.

In this example, the green chiles are the **limiting reactant.** The limiting reactant is the reactant that is consumed first and limits the amount of product that can be made. The same principle applies in determining the quantity of product that can be produced in a chemical reaction. Let's take another look at the reaction of hydrogen and oxygen to produce water, then consider what would happen if 2.00 mol hydrogen and 2.00 mol oxygen were available. How many moles of water can be produced? What is the limiting reactant? Which reactant will be in excess and by how much?

$$2 \text{ H}_2(g) + \text{O}_2(g) \rightarrow 2 \text{ H}_2\text{O}(l)$$

The balanced chemical equation states that 2.00 mol hydrogen react with 1.00 mol oxygen. When the reaction is complete, 2.00 mol water are produced and 1.00 mol oxygen remains unreacted. This problem is easy to solve by inspection. A more systematic way to solve the problem is to create an **SRF** table, as shown in Figure 27. The table is composed of the following lines:

Line 1: The balanced chemical equation is listed.
Line 2: The **S**tarting number of moles of each substance is listed. This would be what is available, the same as the ingredients for the casseroles.
Line 3: The **R**eacting ratio determined from the coefficients in the balanced equation is multiplied by *x*, the basic amount of moles that will react. The reactants are being consumed so a minus sign is placed in front of them. The products are increasing so a plus sign is placed in front of them.
Line 4: This line contains what will be left and what will be formed when the reaction is complete. The values of this line are obtained by adding

lines 2 and 3. This line will provide all the answers, in **Final moles**, to the problems.

		2 H$_2$	+	**O$_2$**	→	**2 H$_2$O**
Line 1						
Line 2	**Starting moles**	2		2		0
Line 3	**Reacting moles**	$-2x$		$-1x$		$+2x$
Line 4	**Final moles**	$2 - 2x$		$2 - 1x$		$0 + 2x$

Figure 27 *Sample SRF Table*

When the reaction is completed, **either** the hydrogen will be consumed **or** the oxygen will be consumed. That means that either

$$2 - 2x = 0 \text{ (hydrogen)} \quad \textbf{or} \quad 2 - 1x = 0 \text{ (oxygen)}$$

If we solve for x in both equations the smallest x value will be the limiting reactant. The larger value will represent an amount in excess of what is possible with the given ingredients.

So for this example, $x = 1$ (hydrogen) or $x = 2$ (oxygen). Since $x = 1$ is smallest, it is the value we choose. This step also identifies the substance that will run out first, the **limiting reactant.** In this case the hydrogen is the limiting reactant.

Using the value of $x = 1$, we can determine the final amounts. The amount of product is $2x = 2(1) = 2$ mol H$_2$O. The reactant in excess is O$_2$ and it is excess by $2 - 1x = 2 - 1(1) = 1$ mol O$_2$.

This method of working the problem gives the same result as the visual inspection—hydrogen is the limiting reactant, and 2.00 mol H$_2$O are produced. The advantage of learning this method is that it works even when the coefficients become difficult to use with the visual method.

Example

Aluminum chloride, AlCl$_3$, has many uses including in deodorants and antiperspirants. It is synthesized from aluminum and chlorine. What mass of AlCl$_3$ can be produced if 100 g of each reactant are available? What is the limiting reactant?

How many grams of the excess reactant remain?

Step 1: Write the balanced equation.
$$2\,Al(s) + 3\,Cl_2(g) \rightarrow 2\,AlCl_3(s)$$

Step 2: Set up the **SRF** Table. Since only moles can go into the table, the grams of each reactant will first need to be converted to moles. See Figure 28.

$$100 \text{ g Al} \times \frac{1 \text{ mol Al}}{27.0 \text{ g Al}} = 3.70 \text{ mol Al}$$

$$100 \text{ g Cl}_2 \times \frac{1 \text{ mol Cl}_2}{71.0 \text{ g Cl}_2} = 1.41 \text{ mol Cl}_2$$

	2 Al(s)	+	3 Cl$_2$(g)	→	2 AlCl$_3$(s)
Starting moles	3.70		1.41		0
Reacting moles	$-2x$		$-3x$		$+2x$
Final moles	$3.70 - 2x$		$1.41 - 3x$		$2x$

Figure 28 *SRF Table*

Step 3: Set the "Final moles" equal to zero and solve for x. Choose the smallest value of x.
$$3.70 - 2x = 0 \text{ or } 1.41 - 3x = 0$$
$$x = 1.85 \text{ mol or } x = 0.47 \text{ mol}$$
Since $x = 0.47$ mol is smaller, Cl$_2$ will be consumed first; it is the limiting reactant.

Step 4: Determine the amount of product formed and the amount of Al remaining by substituting $x = 0.47$ into the corresponding equations on the "Final moles" line.
Amount of product (AlCl$_3$): $2x = 2(0.47 \text{ mol}) = 0.94$ mol AlCl$_3$
Amount of Al remaining: $3.70 - 2(0.47 \text{ mol}) = 2.76$ mol Al

Step 5: Convert the moles back to grams.
$$0.94 \text{ mol AlCl}_3 \rightarrow \frac{134 \text{ g}}{1 \text{ mol}} = 126 \text{ g AlCl}_3 \text{ formed}$$
$$2.76 \text{ mol Al} \rightarrow \frac{27.0 \text{ g}}{1 \text{ mol}} = 74.5 \text{ g Al left}$$

Practice Problems

1. In the synthesis of sodium amide (NaNH$_2$), what is the maximum mass of NaNH$_2$ possible if 50.0 g of Na and 50.0 g NH$_3$ were used?

   $2 \text{ Na(l)} + 2 \text{ NH}_3\text{(g)} \rightarrow 2 \text{ NaNH}_2\text{(s)} + \text{H}_2\text{(g)}$

2. The fuel methanol, CH$_3$OH, can be made directly from carbon monoxide (CO) and hydrogen (H$_2$).
   a. Write a balanced equation for the reaction.
   b. Calculate the maximum mass of methanol if one starts with 5.75 g CO and 10.0 g H$_2$.
   c. Which reactant is the limiting reactant?
   d. How much of the excess reactant remains?

3. Aspirin (C$_9$H$_8$O$_4$) is synthesized in the laboratory from salicylic acid (C$_7$H$_6$O$_3$) and acetic anhydride (C$_4$H$_6$O$_3$):

   $\text{C}_7\text{H}_6\text{O}_3\text{(s)} + \text{C}_4\text{H}_6\text{O}_3\text{(l)} \rightarrow \text{C}_9\text{H}_8\text{O}_4\text{(s)} + \text{CH}_3\text{COOH(l)}$

   a. What is the theoretical yield of aspirin if you started with 15.0 g salicylic acid and 15.0 g acetic anhydride?
   b. Which reactant is the limiting reactant?
   c. What mass of the excess reactant remains?

Appendix A: Teacher Notes

Page 3 Measurement

1 in is a defined quantity and significant figures need not be considered.

Page 7 Precision and Accuracy and Measurement Activity

1. Provide students with as many different balances as possible—either several balances of the same type or, preferably, a variety of types. Examples might include a triple beam, a double pan balance, or various balances that measure to varying levels of accuracy.

2. Select objects that you can identify and label. One suggestion is to use specific heat cylinders made of brass, copper, aluminum, and zinc.

3. Use your most accurate balance to measure the mass of the objects. Define these masses as the "actual" masses of the objects.

Page 9 Significant Figures Activity

1. Provide your students with as many different measuring devices as possible. Include various types of rulers, metersticks, beakers, graduated cylinders, burets, and balances.

2. In the lab set up various stations for students to make measurements.

3. Select one measurement that all the students have recorded. Have each student individually write his or her measurement on a slip of paper and turn it in. Write each measurement on the board. Identify the level of precision of each device, in addition to the range, the average, and the possible errors associated with each.

Page 10 Using Significant Figures Activity

Provide students with a rectangular object and a metric ruler. Expect students to record the correct number of significant digits in their measurements and calculations.

Page 15 Introductory Graphing Activity

It will be helpful to provide students with examples of a wide variety of graphs as they discuss the techniques and rules of graphing. This activity will give students an understanding of graphing concepts and justification for using proper graphing techniques.

Because the students individually create a graph in Step 2, this step is good to assign as homework. Some students may become frustrated by the lack of instruction provided, but assure them that as long as they create a graph, they cannot complete the assignment incorrectly. Remember that the object here is to discover what makes a successful graph.

Students should be divided into groups of 3 or 4 for steps 2–5. After students compare and select the group member's graph that most successfully conveys the table's information, each group should prepare and share a report with the rest of the class. Then a compilation list of all the group lists can be created and used for the duration of the class. The final compilation list should also include the basic graphing rules given on pp. 16.

Page 19 Line Graph Activity

Note that a linear relationship between the number of carbon atoms and the heat of combustion is apparent. This graph enables students not only to see the straight line relationship but also to make predictions that can be tested in the laboratory.

In questions 2 and 3 the students should extrapolate their results by extending their graphs. Explain that the extension should be a dotted line to signify it is only an estimation and that no data was recorded for these values.

Page 30 Weighing Objects to Determine Their Number Activity

This activity should take your students 15 to 20 minutes to complete. Objects that work well are marbles, nails, bolts, nuts, and washers. If different objects have been used, this activity can be extended into having students repeat the process for other objects. A further extension is to list student measurements and have students analyze the data for accuracy, precision, significant figures, and error.

Appendix B: Answer Key

Understanding the Uses of Numbers

Page 3   Practice Problems

1. **a.** meter   **b.** micrometer   **c.** millisecond   **d.** kilogram
       e. nanogram   **f.** milligram   **g.** megagram

2. **a.** kilo   **b.** milli   **c.** centi   **d.** micro

3. $168\ mg\left(\dfrac{1\ g}{1000\ mg}\right) = 0.168\ g$

4. $1.609\ km\left(\dfrac{1000\ m}{1\ km}\right)\left(\dfrac{100\ cm}{1\ m}\right) = 1.609 \times 10^5\ cm$

5. $3.2\ cm\left(\dfrac{10\ mm}{1\ cm}\right) = 32\ mm$

6. The advantage to the SI units is that they are based on multiples of ten. This makes it easy to vary the size of the units by changing the power of ten. Other advantages might include wide international use, common trade pricing, or the flexibility of the system.

Page 4   Practice Problems

1. $340\ L\left(\dfrac{1000\ mL}{1\ L}\right) = 340\ 000\ ml = 3.4 \times 10^5\ mL$

2. $946\ mL\left(\dfrac{1\ L}{1000\ mL}\right) = 0.946\ L$

3. $\dfrac{120\ g}{150\ mL} = 0.80\ \dfrac{g}{mL}$

4. $\dfrac{7.5\ g}{1.9\ cm^3} = 3.9\ \dfrac{g}{cm^3}$

5. $\dfrac{3.93\ g}{5.00\ mL} = 0.786\ \dfrac{g}{mL}$

Page 5   Practice Problems

1. $4\ 741\ 000\ m\left(\dfrac{1\ km}{1000\ m}\right) = 4741\ km$

2. $7265\ ml\left(\dfrac{1\ L}{1000\ mL}\right) = 7.265\ L$

3. $1500\ m\left(\dfrac{39.37\ in}{1\ m}\right)\left(\dfrac{1\ ft}{12\ in}\right)\left(\dfrac{1\ mi}{5280\ ft}\right) = 0.93\ mi$

4. $235\ cm^3\left(\dfrac{2.70\ g}{1\ cm^3}\right) = 635\ g$

5. $2.0\ L\left(\dfrac{1000\ mL}{1\ L}\right)\left(\dfrac{1\ serving}{250\ mL}\right) = 8.0\ servings$

6. $100\ \dfrac{km}{hr}\left(\dfrac{1000\ m}{1\ km}\right)\left(\dfrac{1\ hr}{60\ min}\right)\left(\dfrac{1\ min}{60\ sec}\right) = 27.8\ \dfrac{m}{sec}$

7. $5.4\ L\left(\dfrac{0.17\ g}{1\ L}\right) = 0.92\ g$

8. $7.5\ g\left(\dfrac{1\ mL}{3.12\ g}\right) = 2.4\ mL$

9. $48.5\ mL - 30.0\ mL = 18.50\ mL$ displaced

   $\dfrac{147.8\ g}{18.50\ mL} = 7.99\ \dfrac{g}{mL}$

Page 7   Practice Problems

1. **a.** The sample data is precise and accurate. The average equals 1.34 g/cm³, which is equal to the actual value. The range is only 0.02 g/cm³.
   b. The data is precise (range is only 0.02 g/cm³), but the accuracy is lower. The average is 1.34 g/cm³, but significantly varied from the actual value of 1.40 g/cm³.
   c. The average is 1.37 g/cm³ and the actual value is 1.34 g/cm³, so the accuracy is good. The range is 0.48 g/cm³, so the precision is not very good.
   d. The average is 1.57 g/cm³ and the range is 0.30 g/cm³, so the measurements are neither very accurate or precise.

Page 9–10   Practice Problems

1. **a.** 3   **b.** 4   **c.** 2
   d. 4   **e.** 3

2. **a.** 8.00 cm   **b.** 10.51 cm

3. **a.** 10.5 mL   **b.** 9.0 mL

Page 10–11   Practice Problems

1. **a.** 48.8   **b.** 38.5   **c.** 0.48   **d.** 285   **e.** 119

2. $\dfrac{2.260\ g}{2.04\ mL} = 1.11\ \dfrac{g}{mL}$

3. **a.** 127 cm × 74 cm = 9400 cm² = 0.94 m²
   b. 1.3 m × 0.8 m = 1 m²

   $50.0\ in\left(\dfrac{2.54\ cm}{1\ in}\right) = 127\ cm$

   c. $29.5\ in\left(\dfrac{2.54\ cm}{1\ in}\right) = 74.9\ cm$

   127 cm × 74.9 cm = 9510 cm² = 0.951 m²
   The total area is 3 m² = 0.94 m² + 1 m² + .951 m²

Page 13   Practice Problems

1. **a.** 3.69×10^{-5} **b.** 4.52×10^{-2} **c.** 4.52×10^{6} **d.** 3.6×10^{-2}
   e. 3.65×10^{5}

2. **a.** 1.96×10^{3} **b.** 1.29×10^{-1} **c.** 4.8×10^{1} **d.** 1.2×10^{26}

   e. 5.02×10^{22} **f.** $\left(\dfrac{6.63 \times 10^{-34} J \cdot s \left(3.00 \dfrac{m}{s}\right)}{4.6 \times 10^{-9} m} \right) = 4.3 \times 10^{-25} J$

Understanding and Interpreting Graphs and Tables

Page 14   Introductory Graphing Activity

Students may prepare a combination of graphs that would provide good visualization of the data in the table. Two examples are shown below. Students should point out that two of the plotted points represent four data points. Students should conclude that the dissolved oxygen content is related to water temperature during the year.

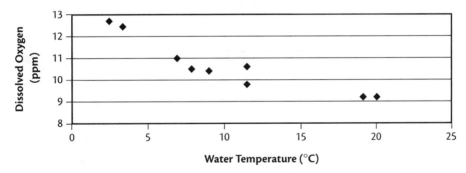

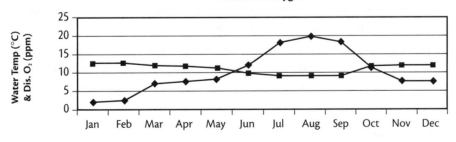

Page 18   Multiple Bar Graphs Activity

1. Within a group, the atomic radius increases as the atomic number increases. This is consistent with all groups shown.

2. As atomic number increases within a period the atomic radius decreases. This is true for all groups shown.

Page 19   Line Graphs Activity

Students may select a variety of graphs to depict the heat of combustion of data. One possible type is shown below. This is an XY-scatterplot that can be used to generate a trendline. The trendline is drawn to best mathematically represent the relationship of the data. From this line interpolations (estimates within the data) and extrapolations (estimates projected beyond the data) can be made.

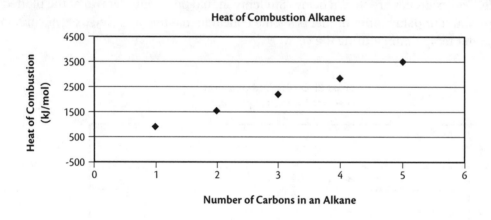

1. As the number of carbon atoms increase, more heat is given off. Students may develop the linear mathematical relationship from the data: $y = m$ (rise/run) $x + b$; where y is the heat of combustion and x is the number of carbons in an alkane, or $y = 654.5\, x + 236.5$.

2. The predicted value of the heat of combustion for the hydrocarbons with 6 carbons would be about 4200 kJ/mole.

3. The heat of combustion would be about 237 kJ/mole. It is not zero because there are still hydrogen atoms left and they could undergo combustion.

4. You could use this to make predictions about other straight-chain hydrocarbons, but it would lose accuracy as the number of carbons grows beyond 5 carbons or the structure of the molecule changes.

Page 20   XY-Plots Activity

1. $\dfrac{rise}{run} = \dfrac{(0.386 - 0.116)\, g}{(0.021 - 0.006)\, ml} = 18.0\, \dfrac{g}{ml}$

2. The appropriate units would be g/ml.

3. The density of gold is 19.3 g/ml, so this value is a little low to be pure gold.

Page 23   Interpreting Tables Activity

1. Answers will vary. The structure and the polarity appear to correlate best with the differences in melting points.

2. Ionic compounds have much higher melting points than do molecular compounds.

3. Molar mass is responsible for the differences in their melting points.

4. Questions will vary. Students could look at relationships between polar and nonpolar molecules. They could also compare numbers of carbon atoms to the melting point.

5. Answers may include boiling point, density, or ratios of characteristics.

Pages 23–25   Additional Graphing Problems Activity

1. Analysis of the CO_2 levels graph:
   a. In 1910, CO_2 level would be about 280 ppm. In 1950 the level would be about 320 ppm; and, in 2000 the level would be about 360 ppm.
   b. An XY-plot would be useful to study this data, in which a mathematical trendline could be drawn to make predictions.

2. The density would be about 1.11 g/ml based on the slope of the graph.

3. a. Answers will vary; most students will choose a line graph with time plotted on the x-axis.
   b. The temperature data indicates a rapid rise in the first three minutes and then a leveling off.
   c. About 92.4 °C
   d. About 101.5 °C
   e. Between 1.5 and 2.0 sec

Understanding Equations and Stoichiometry

Page 29   Balancing Chemical Equations Practice Problems

The coefficients of 1 are used here for emphasis. It is not necessary to include them in the balanced equation.

1. $1 Zn + 2 HCl \rightarrow 1 ZnCl_2 + 1 H_2$

2. $4 Al + 3 O_2 \rightarrow 2 Al_2O_3$

3. $2 C_4H_{10} + 13 O_2 \rightarrow 8 CO_2 + 10 H_2O$

4. $2 KClO_3 \rightarrow 2 KCl + 3 O_2$

5. $3 Fe + 4 H_2O \rightarrow 1 Fe_3O_4 + 4 H_2$

6. $1 CaC_2 + 2 H_2O \rightarrow 1 C_2H_2 + 1 Ca(OH)_2$

7. $1 MnO_2 + 4 HCl \rightarrow 1 MnCl_2 + 2 H_2O + 1 Cl_2$

8. $1 Fe_2O_3 + 3 CO \rightarrow 2 Fe + 3 CO_2$

9. $2 H_2O_2 \rightarrow 2 H_2O + 1 O_2$

10. $1 C_2H_5OH + 3 O_2 \rightarrow 2 CO_2 + 3 H_2O$

Writing and balancing word equations

11. $2\ C_6H_6 + 15\ O_2 \rightarrow 12\ CO_2 + 6\ H_2O$

12. $6\ CO_2 + 6\ H_2O \rightarrow C_6H_{12}O_6 + 6\ O_2$

13. $1\ C_3H_8O_3 + 3\ HNO_3 \rightarrow 1\ C_3H_5N_3O_9 + 3\ H_2O$

14. $1\ Al(OH)_3 + 3\ HCl \rightarrow 1\ AlCl_3 + 3\ H_2O$

15. $3\ NaHCO_3 + 1\ H_3C_6H_5O_7 \rightarrow 3\ CO_2 + 1\ Na_3C_6H_5O_7 + 3\ H_2O$

16. $1\ C_{12}H_{22}O_{11} \rightarrow 11\ H_2O + 12\ C$

Page 31   Moles and Molar Mass Practice Problems

1. 60.05 g/mol

2. 30.03 g/mol

3. 186.33 g/mol

4. 180.16 g/mol

5. 46.07 g/mol

6. 97.99 g/mol

7. 237.93 g/mol

Page 31–32   Mass-Mole Conversions Practice Problems

1. $5.00\ g\ C_7H_6O_3\left(\dfrac{1\ mol}{138.12\ g}\right) = 0.036\ mol$

   $10.5\ g\ CH_3COOH\left(\dfrac{1\ mol}{60.05\ g}\right) = 0.1754\ mol$

2. $500.0\ kg\left(\dfrac{1000\ g}{1\ kg}\right)\left(\dfrac{1\ mol}{186.33\ g}\right) = 2683\ mol$

3. $5.0\ kg\left(\dfrac{1000\ g}{1\ kg}\right)\left(\dfrac{1\ mol}{219.08\ g}\right) = 22.8\ mol =$ approximately 23 *mol*

4. $1.5\ mol\ NaOH\left(\dfrac{40.00\ g}{1\ mol}\right) = 60\ g\ NaOH$

5. $0.123\ mol\ AgNO_3\left(\dfrac{169.88\ g}{1\ mol}\right) = 20.9\ g\ AgNO_3$

Page 32   Calculating Percent Composition Practice Problems

1. % C = 85.6%, % H = 14.4%

2. % C = 85.6%, % H = 14.4%

3. The percent carbon and percent hydrogen are the same in both problems. This is because the ratio of carbon to hydrogen is the same in both compounds.

4. **a.** % Mg = 20.2%, % S = 26.7%, % O = 53.1%

   b. % Mg = 9.9%, % S = 13.0%, % O = 71.4%, % H = 5.7%

   c. % C = 30.7%, % H = 7.75%, % S = 41.0%, % O = 20.5%

   d. % C = 75.9%, % H = 6.4%, % N = 17.7%

   e. % Ca = 38.8%, % P = 20.0%, % O = 41.2%

Page 33   Calculating Percent Compositon Practice Problems

1. % Fe = 69.9%, % O = 30.1%

2. 59% C, 7.1% H, 26.2% O, 7.7% N

3. $24.50 \text{ g} - 15.66 \text{ g} = 8.84 \text{ g } H_2O$

   $\% \, H_2O = \dfrac{8.84 \, g \, H_2O}{24.50 \, g} \times 100 = 36.1\%$

Page 34–35   Formulas from Percent Composition Practice Problems

1. $82.3 \, g \, N \left(\dfrac{1 \, mol \, N}{14.01 \, g \, N} \right) = 5.87 \, mol \, N$

   $17.6 \, g \, H \left(\dfrac{1 \, mol \, H}{1.008 \, g \, H} \right) = 17.5 \, mol \, H$

   divide both by 5.87, then N = 1 mole, H = 3 moles → NH_3

2. $52.2 \, g \, C \left(\dfrac{1 \, mol \, C}{12.01 \, g \, C} \right) = 4.35 \, mol \, C$

   $13.0 \, g \, H \left(\dfrac{1 \, mol \, H}{1.008 \, g \, H} \right) = 12.90 \, mol \, H$

   $34.8 \, g \, O \left(\dfrac{1 \, mol \, O}{16.00 \, g \, O} \right) = 2.175 \, mol \, O$

   divide all by 2.175 → C_2H_6O

3. $74.03 \, g \, C \left(\dfrac{1 \, mol \, C}{12.01 \, g \, C} \right) = 6.16 \, mol \, C$

   $8.70 \, g \, H \left(\dfrac{1 \, mol \, H}{1.008 \, g \, H} \right) = 8.63 \, mol \, H$

   $17.27 \, g \, N \left(\dfrac{1 \, mol \, N}{14.01 \, g \, N} \right) = 1.23 \, mol \, N$

   divide all by 1.23 → C_5H_7N

4. $82.66 \, g \, C \left(\dfrac{1 \, mol \, C}{12.01 \, g \, C} \right) = 6.882 \, mol \, C$

   $17.34 \, g \, H \left(\dfrac{1 \, mol \, H}{1.008 \, g \, H} \right) = 17.20 \, mol \, H$

   divide both by 6.882; giving C = 1, H = 2.5, then multiply by 2 → C_2H_5

5. $.252 \, g \, Ti \left(\dfrac{1 \, mol \, Ti}{47.87 \, g \, Ti} \right) = .00526 \, mol \, Ti$

   $.748 \, g \, Cl \left(\dfrac{1 \, mol \, Cl}{35.45 \, g \, Cl} \right) = .0211 \, mol \, Cl$

   divide both by .00526 → $TiCl_4$

Page 36   Understanding Mass Relationships Practice Problems

1. $25.0 \, mol \, C_2H_2 \left(\dfrac{4 \, mol \, CO_2}{5 \, mol \, C_2H_2} \right) = 20 \, mol \, CO_2$

2. $5.0 \, g \, C_4H_{10} \left(\dfrac{1 \, mol \, C_4H_{10}}{58 \, g \, C_4H_{10}} \right) \left(\dfrac{8 \, mol \, CO}{9 \, mol \, C_4H_{10}} \right) \left(\dfrac{28 \, g \, CO}{1 \, mol \, CO} \right) = 2.1 \, g \, CO$

3. $15.0 \ g \ NaNH_2\left(\dfrac{1 \ mol \ NaNH_2}{39.0 \ g \ NaNH_2}\right)\left(\dfrac{2 \ mol \ Na}{2 \ mol \ NaNH_2}\right)\left(\dfrac{23.0 \ g \ Na}{1 \ mol \ Na}\right) = 8.85 \ g \ Na$

4. $100 \ g \ CH_3COOC_2H_5\left(\dfrac{1 \ mol \ CH_3COOC_2H_5}{88.1 \ g \ CH_3COOC_2H_5}\right)\left(\dfrac{1 \ mol \ CH_3COOH}{1 \ mol \ CH_3COOC_2H_5}\right)$

$\left(\dfrac{60.0 \ g \ CH_3COOH}{1 \ mol \ CH_3COOH}\right) = 68.2 \ g \ CH_3COOH \left(\dfrac{100 \ g \ total}{35 \ g \ used}\right) = 195 \ g \ CH_3COOH$

5. $CaCO_3 \rightarrow CaO + CO_2$

$4.65 \ g \ CaO\left(\dfrac{1 \ mol \ CaO}{56.1 \ g \ CaO}\right)\left(\dfrac{1 \ mol \ CaCO_3}{1 \ mol \ CaO}\right)\left(\dfrac{100.1 \ g \ CaCO_3}{1 \ mol \ CACO_3}\right) = 8.30 \ g \ CaCO_3$

Page 39   Understanding Limiting Reactants Practice Problems

1. $50 \ g \ Na\left(\dfrac{1 \ mol}{22.99 \ g}\right) = 2.17 \ mol \ Na$ $\qquad$ $50 \ g \ NH_3\left(\dfrac{1 \ mol}{17.03 \ g}\right) = 2.94 \ mol \ NH_3$

	2 Na(l)	+	2 NH$_3$(g)	→	2 NaNH$_2$(s)	+	1 H$_2$(g)
	2.17		2.94		0		0
Reacting moles	$-2x$		$-2x$		$+2x$		$+1x$
Final moles	$2.17 - 2x$		$2.94 - 2x$		$2x$		$1x$

Then, either $2.17 - 2x = 0$ or, $2.94 - 2x = 0$

$x = 1.085$ or $x = 1.47$; Since $x = 1.085$ is smaller, it is the correct value, and sodium is the limiting reactant. Therefore, the maximum amount of $NaNH_2$ that can be produced is

$2.17 \ mol \ Na\left(\dfrac{2 \ mol \ NaNH_2}{2 \ mol \ Na}\right)\left(\dfrac{39.02 \ g \ NaNH_2}{1 \ mol \ NaNH_2}\right) = 84.7 \ g \ NaNH_2$

2. $CO(g) + 2 \ H_2(g) \rightarrow CH_3OH(l)$

a. $5.75 \ g \ CO\left(\dfrac{1 \ mol}{28.0 \ g}\right) = 0.205 \ mol \ CO$ $\qquad$ $10.0 \ g \ H_2\left(\dfrac{1 \ mol}{2.02 \ g}\right) = 4.95 \ mole \ H_2$

	CO(g)	+	2 H$_2$(g)	→	CH$_3$OH(l)
Starting moles	0.205		4.95		0
Reacting moles	$-x$		$-2x$		$+x$
Final moles	$0.205 - x$		$4.95 - 2x$		x

$x = 0.205$

b. Maximum amount of CH_3OH is x = 0.205 mol = 6.57 g

c. CO is the limiting reactant.

d. Excess H_2 = 4.95 moles − 2(0.205 mol) = 4.54 mol = 9.17 g H_2

3. $15.0 \text{ g } C_7H_6O_3\left(\dfrac{1 \text{ mol}}{138.1 \text{ g}}\right) = 0.109 \text{ mol salicylic acid}$

$15.0 \text{ g } C_4H_6O_3\left(\dfrac{1 \text{ mol}}{102.1 \text{ g}}\right) = 0.147 \text{ mol acetic anhydride}$

	$C_7H_6O_3$	+	$C_4H_6O_3$	$\rightarrow$	$C_9H_8O_4$	+	CH_3COOH
Starting moles	0.109		0.147		0		0
Reacting moles	$-x$		$-x$		$+x$		$+x$
Final moles	$0.109 - x$		$0.147 - x$		x		x

a. The theoretical yield of aspirin $= x = 0.109 \text{ mol} = 19.5 \text{ g}$

b. The limiting reactant is salicylic acid, $C_7H_6O_3$

c. The excess reactant is acetic anhydride.

The excess equals $(0.147 - x) \text{ mol} = (0.147 - 0.109) = 0.038 \text{ mol} = 3.92 \text{ g}$.